All Kinds of Sounds

by Maryellen Gregoire

Consultant:
Adria F. Klein, Ph.D.
California State University, San Bernardino

capstone
classroom

Heinemann Raintree • Red Brick Learning
division of Capstone

This is a loud sound.

This is a soft sound.

This is a loud sound.

This is a soft sound.

This is a loud sound.

This is a soft sound.

We all make sounds.